Maths Revision Booklet
for CCEA GCSE 2-tier specification
M5

Conor McGurk

Contents

Revision Exercise 1A (non-calculator) 3

Revision Exercise 1B (non-calculator) 10

Revision Exercise 2A (with calculator) 17

Revision Exercise 2B (with calculator) 23

Problem Solving Questions. 29

Answers 31

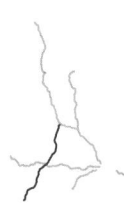

# Revision Exercise 1A (non-calculator)

You must **not** use a calculator for this paper. Total mark for this paper is 50.
Figures in brackets printed down the right-hand side of pages indicate the marks awarded to each question or part question.
You should have a ruler, compasses, set-square and protractor.

1. The temperature is shown on the thermometer below in both °F and °C.

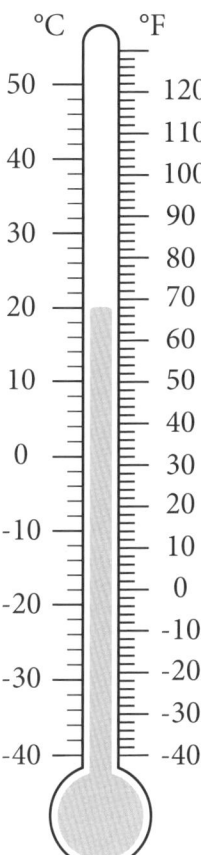

(a) What temperature in °F is 20°C?

Answer _____ °F [1]

(b) What temperature in °C is 44°F?

Answer _____ °C [1]

2. **Impossible   Unlikely   Evens   Likely   Certain**

   Choose from the words above to describe the probability of the following events happening.

   (a) A fair dice will show an even number.

   Answer _____ [1]

   (b) Tomorrow will be a day ending in the letter "Y".

   Answer _____ [1]

   (c) Your best friend will meet Ed Sheeran next week.

   Answer _____ [1]

   (d) A person selected from your school is right handed.

   Answer _____ [1]

3. (a) John and his 3 friends are going to "Fry Days" to get some carry out food.
   John wants a burger and chips which costs £3.85
   His friends want a fish supper at £4.15, sausage and chips at £3.15 and chicken and chips at £4.95
   Estimate the total cost of the meals.
   Show all your working.

   Answer £ _____ [2]

   (b) Michael estimates that the square root of 80 is about 8
   Is Michael correct?
   Explain your answer.

   Answer _____ because _____ [2]

4. Put the four digits below in order to make a number as close as possible to 4 thousand.

   9   4   3   5

   Answer _____ [2]

5. I have 8 coins and 15p in total. I have at least one of each of the coins 5 pence, 2 pence and 1 pence.
   How many of each coin do I have?

   Answer _____ 1p _____ 2p _____ 5p [3]

Revision Exercise 1A

6. A person is to be selected at random from a youth club to represent the club at an international event.

(a) Explain why the probability of selecting a girl may not be ½

Answer _____ [1]

There are 24 girls and 30 boys in the youth club.

(b) What is the probability of selecting a boy at random from the club?
Give your answer in its simplest form.

Answer _____ [2]

7. (a) A sack of potatoes weighs 25 kg. What is the approximate weight of the sack of potatoes in pounds?

Answer _____ pounds [2]

(b) The distance between Belfast and Dublin is 100 miles. How many kilometres is this distance?

Answer _____ km [2]

8. (a) A bag contains 4 blue balls, 5 red balls and 6 green balls.
What is the probability of choosing at random a red ball?

Answer _____ [2]

9. Here are the first three patterns in a sequence.

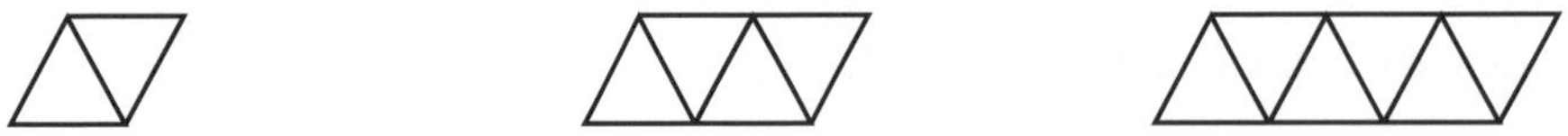

Pattern number 1   Pattern number 2   Pattern number 3

(a) How many triangles are there in pattern 7?

Answer _____ [2]

(b) There are 30 triangles in a pattern number.
What pattern number is it?

Answer _____ [2]

(c) Tom says that there can be no pattern number with 51 triangles.
Is Tom correct? Give a reason for your answer.

Answer _____ because _____ [2]

10. Draw a right angled triangle with sides making up the right angle 5 cm and 12 cm long.
Measure the third side of the triangle.

Answer _____ cm [3]

11. Two coins are tossed. The outcomes are entered in the table.

Coin	Head (H)	Tail (T)
Head (H)	(H, H)	
Tail (T)		

Complete the possible outcomes in the table.

[1]

What is the probability of 2 heads occurring?

Answer _____ [1]

12. Look at the shape below.

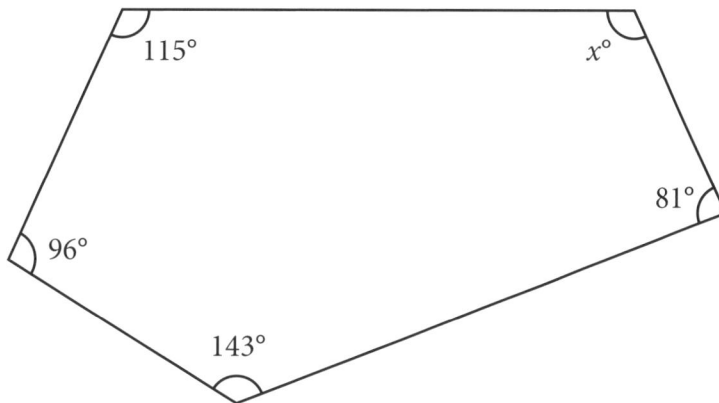

Find the value of x.

Answer $x =$ _____ ° [3]

13. Lisa leaves home for work at 0710. Her journey is shown on the graph below. Lisa's brother, Matthew, leaves home at 0700 and travels for half an hour at an average speed of 80 km/hr. He stops for 5 minutes to buy a newspaper. Then, Matthew continues on his journey at an average speed of 60 km/hr and arrives at his work at 0810.

   (a) Draw Matthew's journey on the graph below.

   [2]

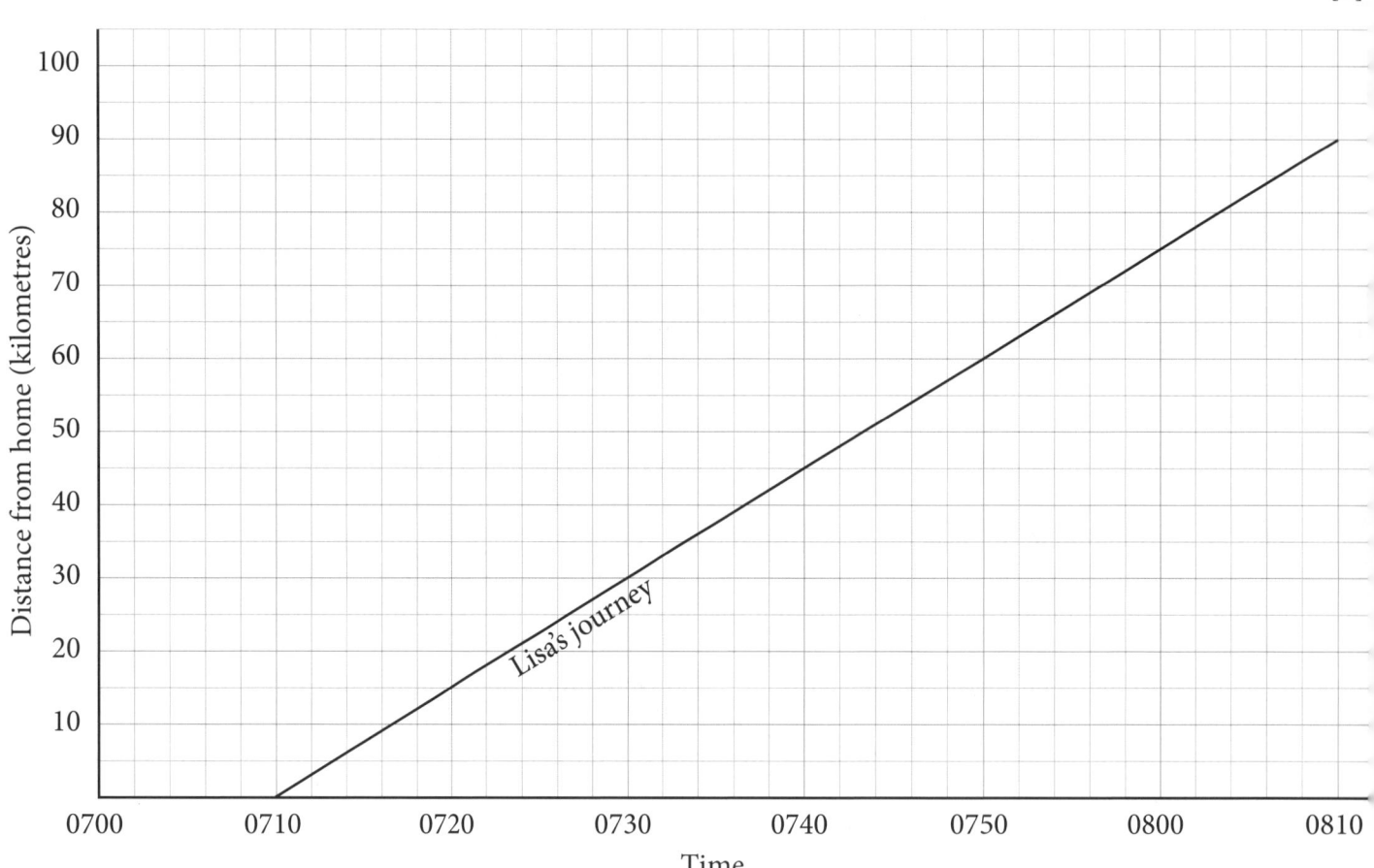

   (b) At what time does Lisa pass Matthew on the road?

   Answer _____ [1]

   (c) How far is Matthew from his work at the time Lisa overtakes him?

   Answer _____ [1]

14. At a wedding reception, one half of those attending ate roast beef only, one third ate turkey only and 16 people ate salmon only. Everyone attending the wedding ate roast beef or turkey or salmon.

   How many people were at the wedding reception?

   Answer _____ [4]

15. The graph below shows a shape, A.

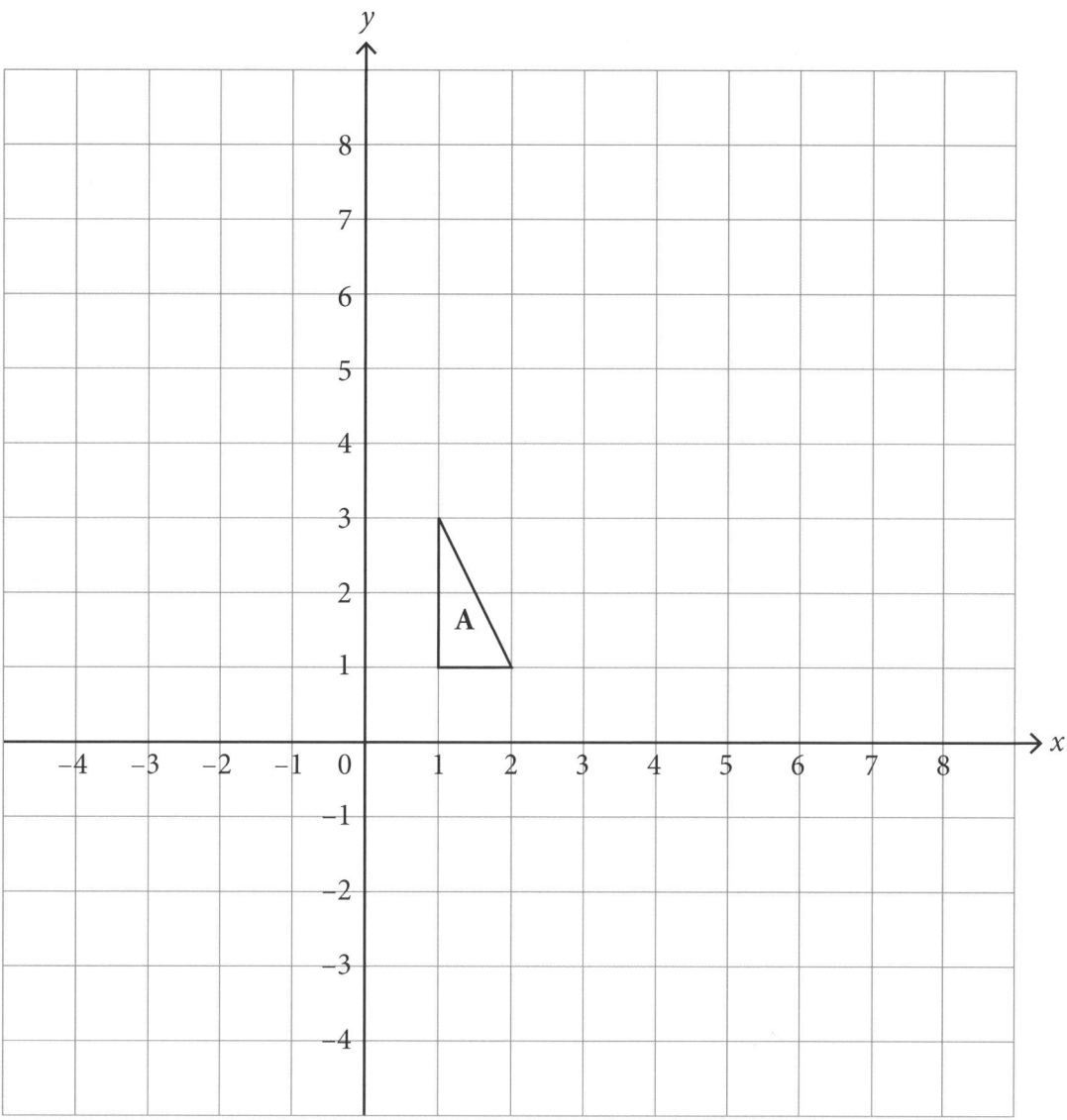

(a) Reflect shape A in the *y*-axis.
Label the shape **B**.

[1]

(b) Enlarge shape A by a scale factor 2 from the centre (−1 , 0).

[3]

Revision Exercise 1B

(non-calculator)

1. Classify the following events as **certain**, **likely**, **evens**, **unlikely** or **impossible**.

   (a) Belfast will have 365 days of unbroken sunshine in one year.

   Answer _____ [1]

   (b) March will have 31 days next year.

   Answer _____ [1]

   (c) I toss a coin and get a head.

   Answer _____ [1]

   (d) I toss a fair dice and get a 6

   Answer _____ [1]

2. (a) How many lines of symmetry has the shape below?

   Answer _____ [1]

   (b) Reflect the shape shown below in the mirror line AB.

   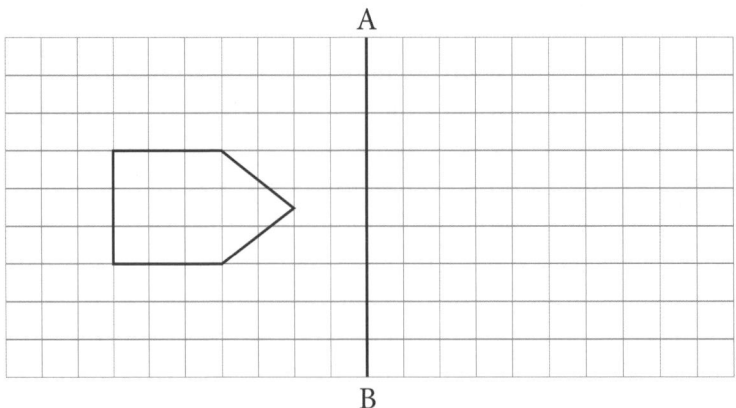

   [2]

   (c) What is the order of rotational symmetry of a parallelogram?

   Answer _____ [1]

15. The graph below shows a shape, A.

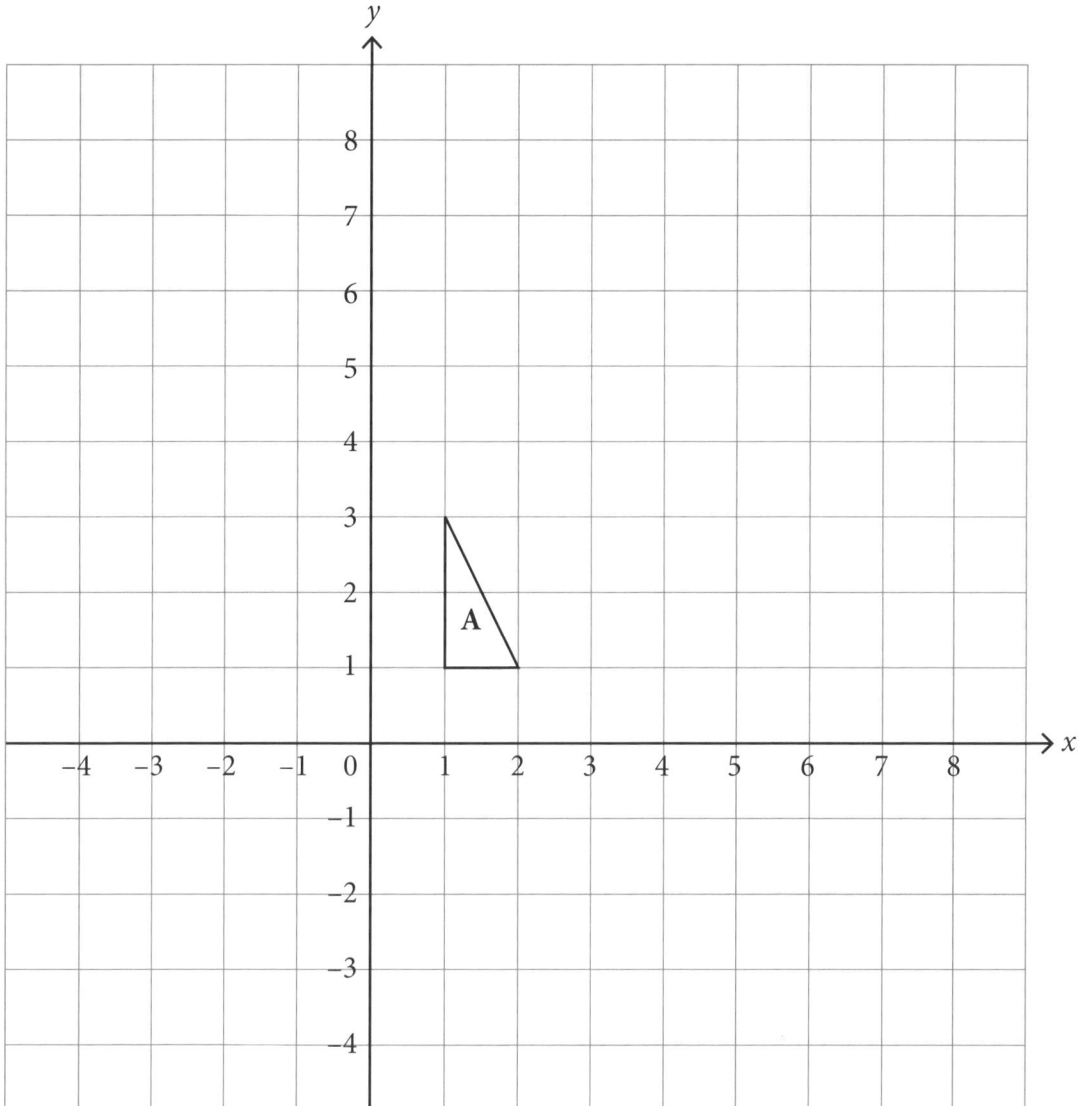

(a) Reflect shape A in the y-axis.
    Label the shape **B**.

[1]

(b) Enlarge shape A by a scale factor 2 from the centre (−1 , 0).

[3]

Revision Exercise 1B

(non-calculator)

1. Classify the following events as **certain**, **likely**, **evens**, **unlikely** or **impossible**.

   (a) Belfast will have 365 days of unbroken sunshine in one year.

   Answer _____ [1]

   (b) March will have 31 days next year.

   Answer _____ [1]

   (c) I toss a coin and get a head.

   Answer _____ [1]

   (d) I toss a fair dice and get a 6

   Answer _____ [1]

2. (a) How many lines of symmetry has the shape below?

   Answer _____ [1]

   (b) Reflect the shape shown below in the mirror line AB.

   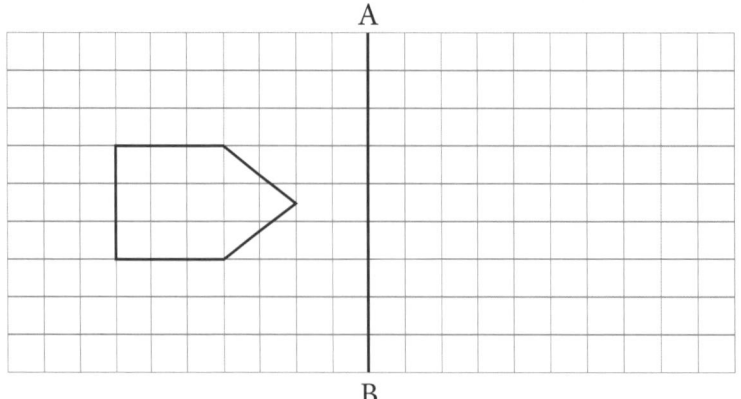

   [2]

   (c) What is the order of rotational symmetry of a parallelogram?

   Answer _____ [1]

Revision Exercise 1B

3. What change will I get from £5 after buying 4 apples at 49p each?

Answer _____ [2]

4. The markings on the probability scale below are at equal intervals.

0 |————|————|————|————|————|————| 1

Mark the point on the scale:

(a) with an **X** the probability of getting a head when a coin is tossed. [1]

(b) with a **Y** the probability of the name of a month starting with the letter Q. [1]

(c) with a **Z** the probability of getting a 6 on a fair dice. [1]

5. (a) Estimate 7.9×97.8

Answer _____ [2]

(b) Estimate how many boxes of cereal costing £2.85 can be bought for £27

Answer _____ [2]

6. (a) The probability that a package posted on Friday will be delivered on Saturday is 0.8
What is the probability that it is not delivered?

Answer _____ [2]

(b) Write your answer to part (a) as a fraction in its simplest form.

Answer _____ [2]

7. Look at the shape below.

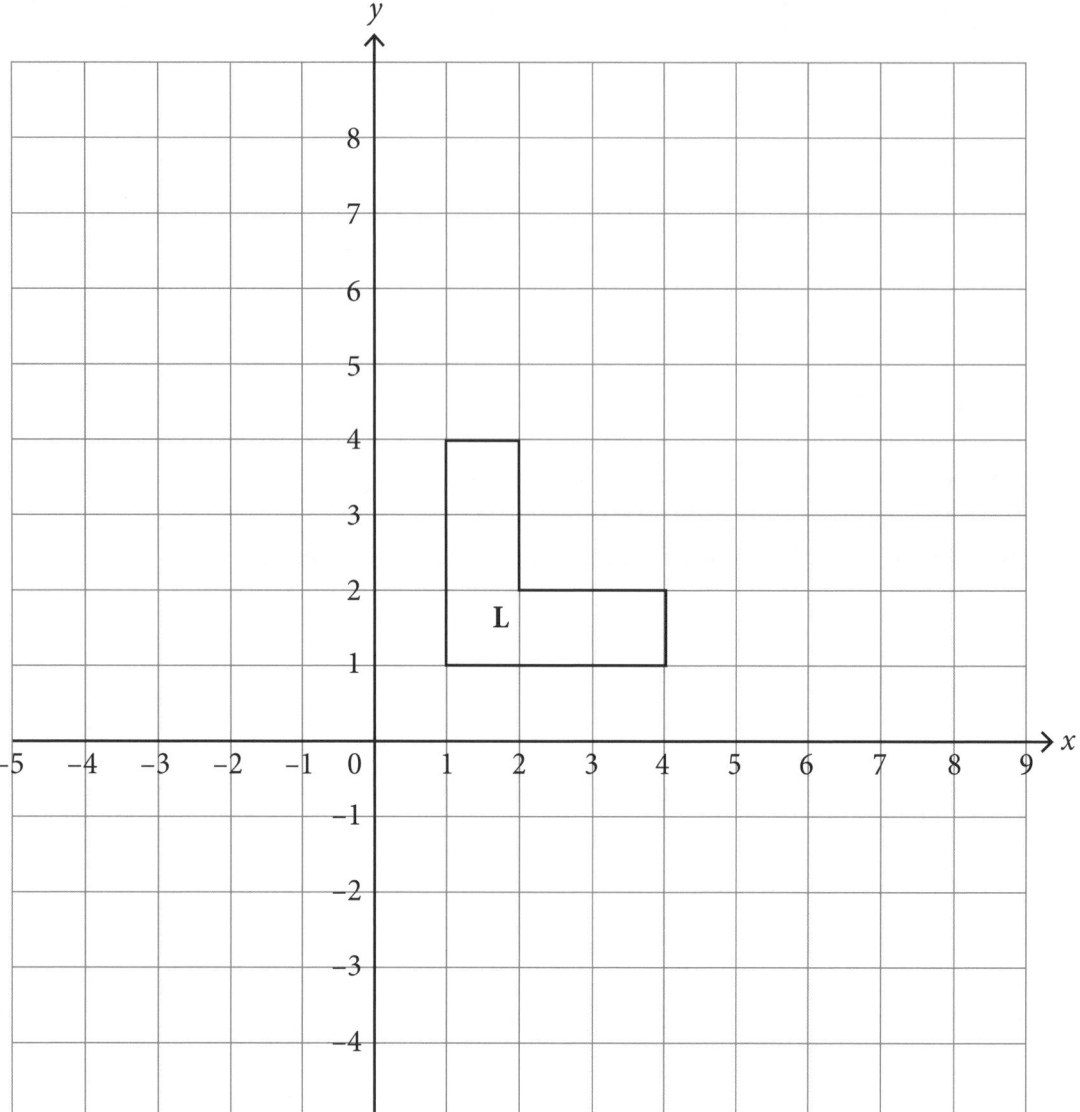

(a) How many lines of symmetry has shape L?

Answer _____ [1]

(b) Translate the shape 4 right, 4 down.
Label it **M**.

Answer _____ [1]

(c) Draw the image of L after a reflection in the *x*-axis.
Label it **N**.

Answer _____ [2]

8. Three spinners A, B and C are shown below.

A B C

(a) What is the likelihood of any of the spinners landing on a 4?

Answer _____ [1]

(b) Which spinner is most likely to land on a 2?

Answer Spinner _____ [1]

(c) Which spinner would you choose if you wanted the spinner to land on a 1?

Answer Spinner _____ [1]

(d) Explain why there is an even chance that spinner C will land on a 3.

Answer _____ [1]

9. Bicycles can be hired at £5 for the first hour and £2 per hour for each hour after that.
John and Elizabeth wish to hire a bicycle each.
They have £25 between them.
For how many hours can they hire the bicycles?

Answer _____ [4]

10. Work out:

    ⅓ + ¾

    Answer _____ [2]

11. How many ways are there of arranging the letters P, Q and R in a straight line?
    The first 2 have been done for you.

    PQR
    PRQ

    Answer _____ [2]

12. Sara knows that it takes ¾ of a tin of baby food to feed her child each day.
    How many tins does she need to buy to last for a full week?

    Answer _____ tins [3]

13. In a game of chance, the probability that a contestant wins is ¼
    If 300 contestants take part, how many would you would expect to win?

    Answer _____ [2]

14. The conversion graph below may be used to convert kilograms into pounds and vice versa.

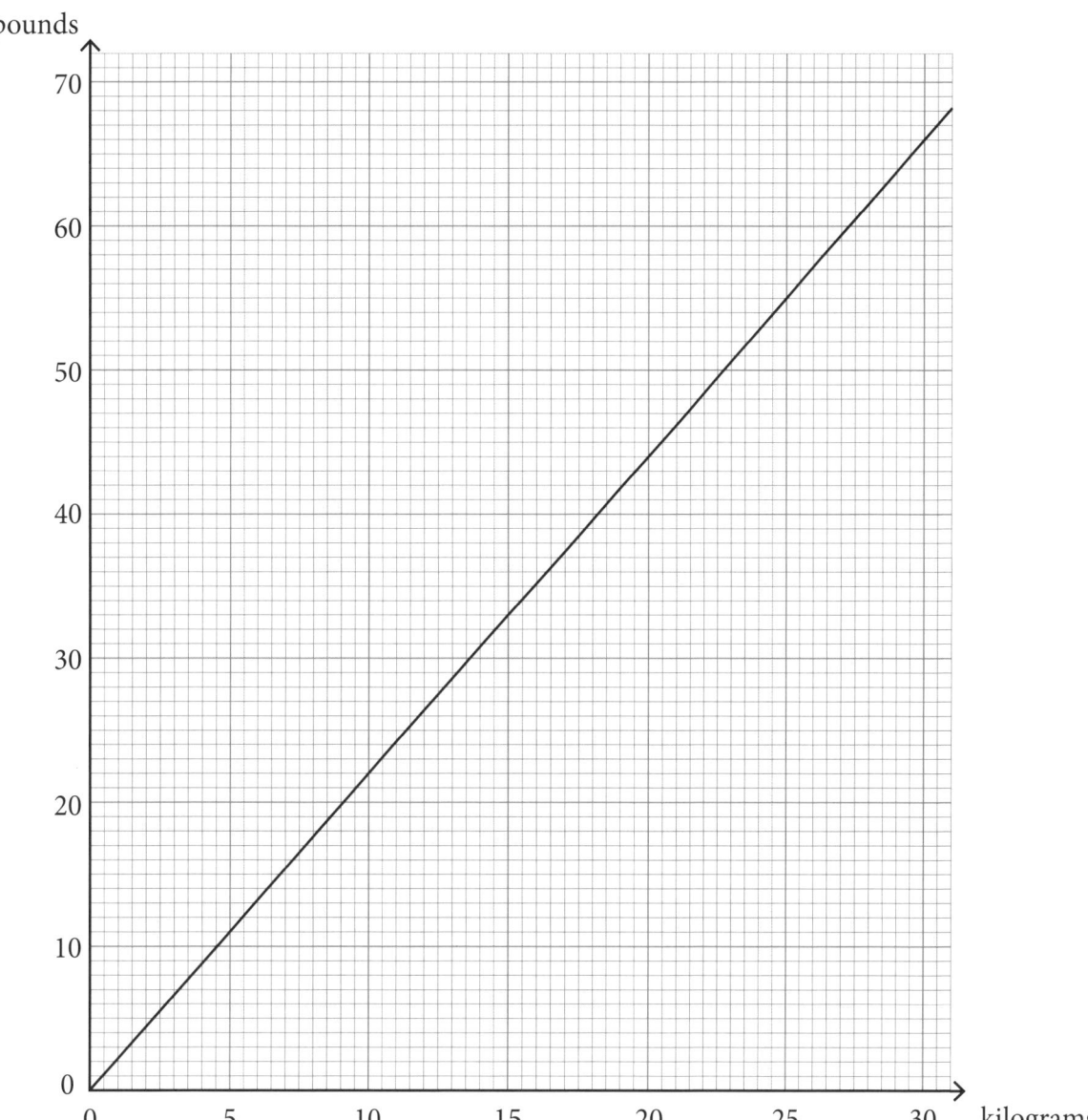

Use the graph to convert:

(a) 12 kilograms into pounds.

Answer _____ pounds [1]

(b) 47 pounds into kilograms.

Answer _____ kilograms [1]

(c) 440 pounds into kilograms.

Answer _____ kilograms [1]

15. Given that:

    75.2 × 53.5 = 4023.2

    find the value of :

    (a) 7.52 × 5.35

    Answer _____ [1]

    (b) $\frac{4023.2}{0.752}$

    Answer _____ [1]

16. The diagram below shows two shapes, A and B.

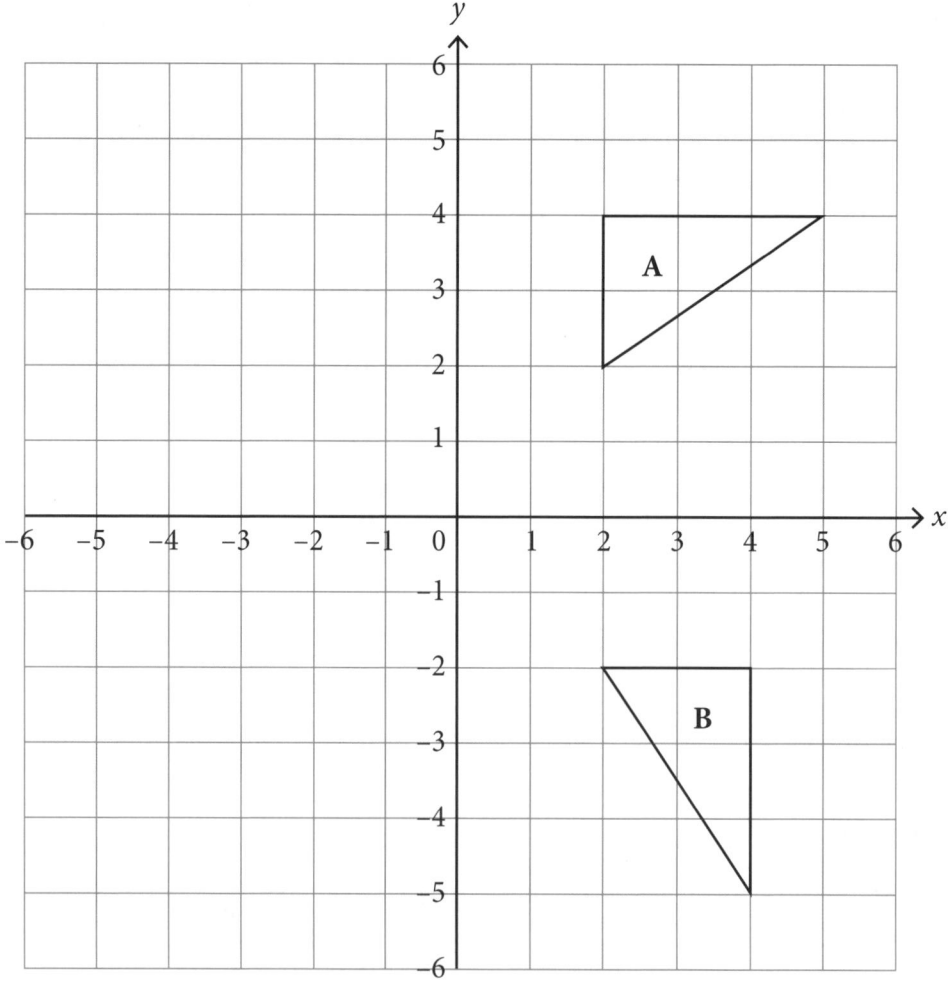

Describe fully the single transformation that maps triangle **A** onto triangle **B**.

Answer _____

_____ [2]

Revision Exercise 2A (with calculator)

1. Draw the reflection of the shape in the mirror line.

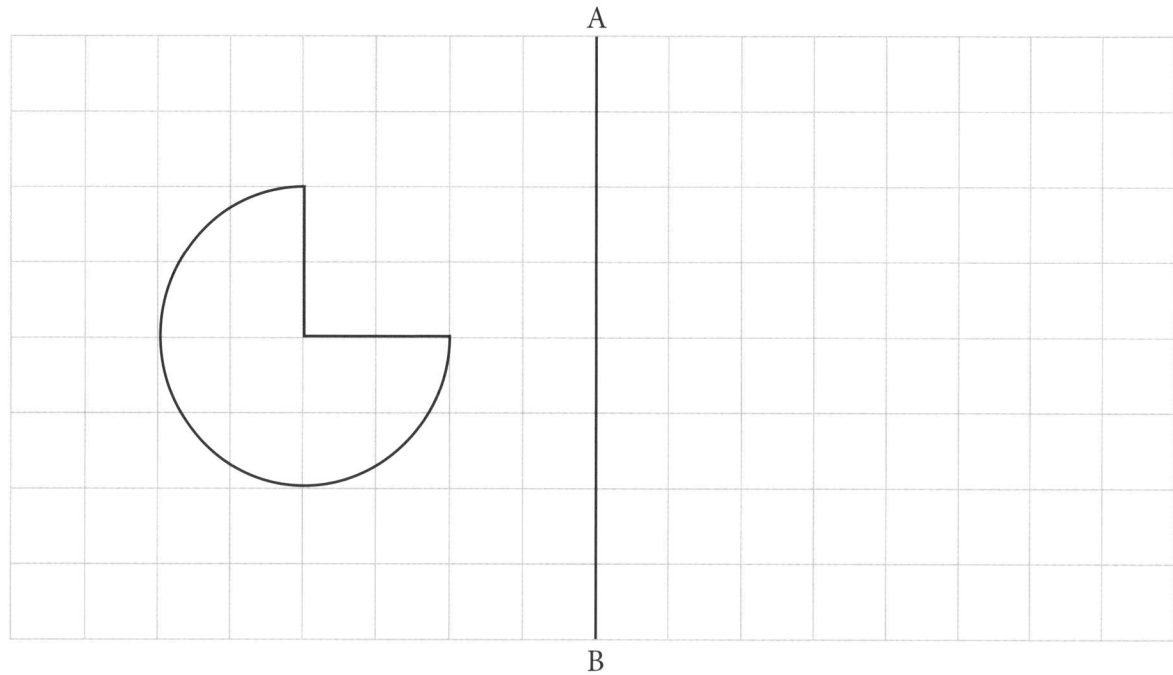

[1]

2. Shona buys 1.5 kg of grapes at £4.20 per kg.
   She also buys 3 cartons of orange juice at £2.45

   (a) How much change does she receive from a £20 note?

   Answer _____ [3]

   (b) She receives 10 reward points for each complete pound she spends.
       How many reward points does she receive?

   Answer _____ [1]

3. Choose a word from the list which matches the possibility of the events below happening.

impossible   unlikely   evens   likely   certain

(a) Selecting a day of the week beginning with the letter F.

Answer _____ [1]

(b) Obtaining a 12 when a fair dice is thrown.

Answer _____ [1]

(c) Getting a head or a tail when a coin is tossed.

Answer _____ [1]

(d) Selecting an even number or a prime number from a bag of balls labelled from 1 to 10

Answer _____ [1]

4. (a) Add **one** square to the shape below so that the shape has one line of symmetry.

[1]

(b) Add **two** squares to the shape below so that the shape has one line of symmetry.

[1]

(c) Add **one** square to the shape below so that the shape has rotational symmetry of order 2.

[1]

5. The *Termocycle* team are preparing a circuit for their annual cycle competition. The distances between stopping points have all been calculated in miles. The team wish to convert the distances to kilometres using the conversion 5 miles = 8 kilometres.

(a) Draw a conversion graph on the grid below.

[2]

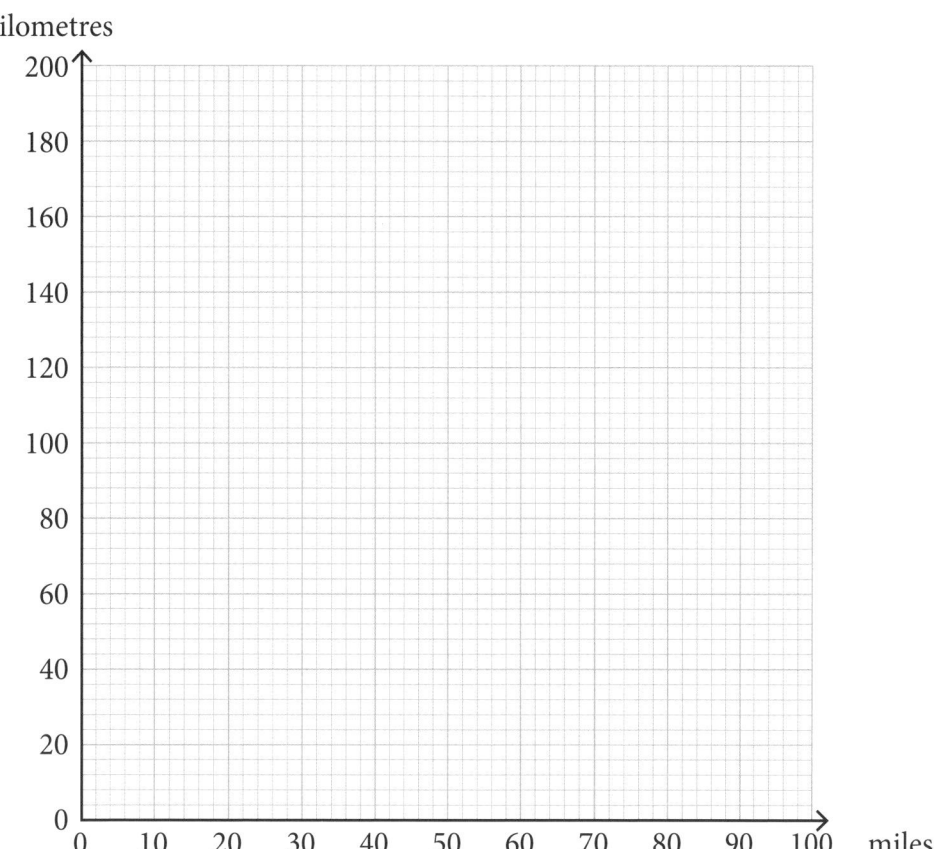

(b) What is the distance in kilometres of 35 miles?

Answer _____ km [1]

(c) Two stopping points are 120 miles apart.
Explain how you can use the graph to work out this distance in kilometres.

Answer _____

_____

_____ [2]

6. **AppDeals** is a new app that provides Bell Phone users with daily deals straight from developers. Now, Bell Phone users can get discounted apps or free apps every day as given below.

What is the best price I can pay for 6 apps?
Show your working.

Answer _____ [5]

7. An experiment consists of spinning a 5 sided spinner with the numbers 1 to 5 and tossing a coin. List all the possible outcomes in the table below.

	Spinner				
Coin	(1, H)				

[2]

8. **(a)** What four-sided shape has rotational symmetry of order 4?

Answer _____ [1]

(b) Calculate the size of angle x in the diagram below.

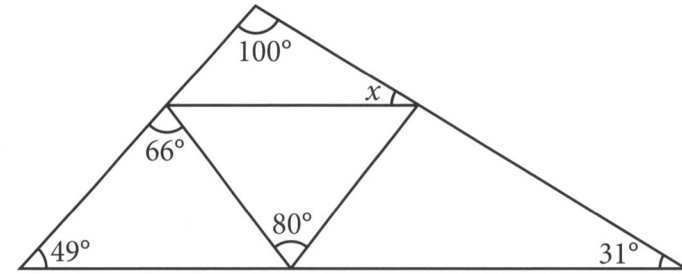

Answer _____ ° [3]

9. 6 inches is approximately 15 cm.
   1 yard is 36 inches.

   Approximately, how many **metres** are there in 10 yards?

   Answer _____ m [3]

10. Mary has a combination lock to prevent her bicycle being stolen.
    The code for the lock consists of 4 digits.
    The four digits she uses are 2, 4, 6 and 8.

    (a) How many different combinations are there with these 4 digits?

    Answer _____ [2]

    (b) Mary has forgotten her code, but can remember that the second digit is 4
        How many different combinations are there that Mary can try with 4 as the second digit in order to be able to open the combination lock?

    Answer _____ [2]

11. ABCD, shown below, is a rhombus.

    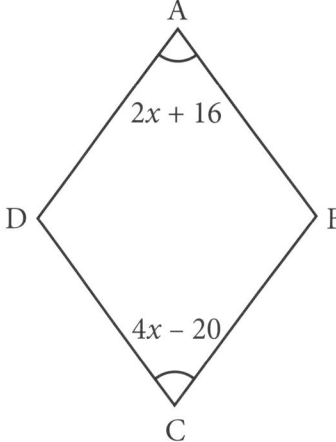

    (a) Find the size of the angle at A.

    Answer _____ ° [3]

    (b) Hence, find the size of angle at B.

    Answer _____ ° [3]

12. Pupils in a school were asked about their method of transport to school.
The table below shows the probability of a pupil using the different modes of transport.

Method of transport	Bus	Car	Cycle	Walk
Probability	0.35	0.25		0.23

(a) Given that all the pupils travel to school by one of the methods in the table, work out the probability that a pupil chosen at random cycles to school.

Answer _____ [2]

(b) If there are 600 pupils at the school, how many cycle to school?

Answer _____ [2]

13. John, Mary and Elizabeth share an amount of money in the ratio 2:3:5
Elizabeth received £24 more than Mary.
Calculate the total amount of money.

Answer _____ [5]

Revision Exercise 2B (with calculator)

1. A number is selected at random from the first four even numbers.
   Another number is selected at random from the factors of six.
   The two numbers are added.

   (a) Complete the table below, filling in the missing values.

   Even numbers

	2		6	8
Factors of six					
6				

   [4]

   Using the table above, what is the probability that the sum of the two numbers is:

   (b) an even number?

   Answer _____ [1]

   (c) a prime number?

   Answer _____ [2]

2. There are four sweets in a bag.
   There are three jelly sweets and one mint sweet.

   (a) What is the probability that a sweet chosen at random from the bag is a chocolate sweet?

   Answer _____ [1]

   (b) What is the probability that a sweet chosen at random from the bag is a jelly sweet?

   Answer _____ [1]

3. William is going on holiday to Majorca.
   The airline permits a baggage allowance of 20 kilograms.
   William weighs his suitcase and his scales show that it has a weight of 48 pounds.

   (a) Is William's suitcase too heavy?
   Show all your working.

   Answer _____ , because _____ [3]

   (b) William takes a number of books from his suitcase and puts them in his wife's luggage.
   He is told that the weight of the suitcase is 20 kg to the nearest kilogram.
   What are the greatest and least values of the weight of his suitcase?

   Answer: Greatest _____ kg and least _____ kg [2]

4. A shop sells packs of cereal at £2.80 each. On one Saturday, they have a special offer on the cereal:

   "Buy 2 packs and get a third pack FREE!"

   The following Saturday, the offer changes to:

   "Buy one pack and get a second pack HALF price!"

   Which is the best offer, the first or the second?
   You must show all your working and explain your answer.

   Answer _____
   _____ [4]

5. Enlarge shape A by a scale factor 2 about the origin.

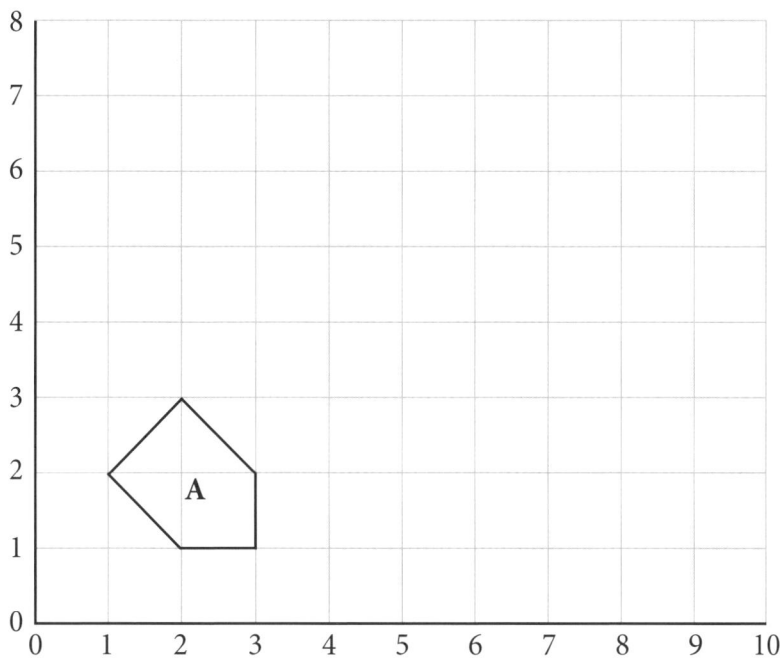

[3]

6. John's code to enter a secure room consists of the digits 8, 0 and 6 and another missing digit.

8 0 ☐ 6

Any of the digits from 0 to 9 may be used.
John has forgotten the third digit of his code but he knows that he has not repeated any digits.

(a) What is the probability that he picks an even number?

Answer _____ [1]

John's sister tells John that the missing digit is odd but is not the digit 1

(b) What is the probability that John picks the correct digit?

Answer _____ [1]

7. A car costs 2.4 million yen in Japan.
If £1 = 150.79 yen, calculate the cost of the same car in £ sterling.
Give your answer to the nearest £100.

Answer £ _____ [3]

8. **(a)** The rule for creating a sequence is:

    To get the next term subtract 4 from the previous term.

    The first term = 15

    Write down the second, third and fourth terms in the sequence.

    Answer 15, _____ , _____ , _____ [2]

   (b) The rule for creating another sequence is:

    To get the next term multiply the previous term by 3 and subtract 4

    The first term = 6

    Write down the second, third and fourth terms in the sequence.

    Answer 6, _____ , _____ , _____ [2]

9. To make dark green paint, blue and yellow paint are mixed in the ratio 2:1
   Blue paint is £15 per litre.
   Yellow paint is £12 per litre.

   Matthew requires 10 litres of dark green paint.

   What is the minimum cost to make the 10 litres if the paint is only sold in 1 litre cans?

   Answer £ _____ [5]

10. Make an accurate drawing of a rhombus of side 8 cm.
The shorter diagonal of the rhombus should be 6 cm.

[4]

11. Jonathan drives to work each morning. He leaves home at 0655 and returns home in the evening at 1635
   (a) Work out how long Jonathan is away from home.

Answer _____ [1]

Jonathan's workplace is 25 kilometres from his home.
He takes 20 minutes to travel to his workplace.

(b) What is his average speed in kilometres per hour?

Answer _____ [2]

(c) Approximately how many miles is Jonathan's house from his workplace?
Give your answer correct to the nearest tenth of a mile.

Answer _____ miles [1]

12. A recipe for vanilla ice cream for 4 people is given below.

Vanilla Ice Cream

600 ml double cream

125 g icing sugar

2 teaspoons vanilla extract

2 eggs

A chef has 1500 ml of double cream, 350 g of icing sugar, 8 teaspoons of vanilla extract and 6 eggs.

For how many people can she make ice cream?

Show all your working.

Answer _____ people [5]

Problem Solving Questions

Note: *The new CCEA GCSE in Mathematics has an increased weighting for problem solving tasks: 25% AO3 for Foundation tier and 30% for Higher tier.*

A problem solving question is one where the student will, most likely, not see an immediate method for solving it. Hence, the student should persevere with the problem using a range of strategies. This section contains some problem solving questions to help students to practice working in this way. See CCEA's GCSE Mathematics microsite for further problem solving examples.

1. There is a special offer on football cards in a shop. Each card comes with a "special voucher". Once you have collected 4 vouchers you can exchange them for another free card. If you buy 32 football cards, how many free cards in total can you get?

   Answer _____ [3]

2. The diagram below consists of 16 small squares of side 1 cm.

   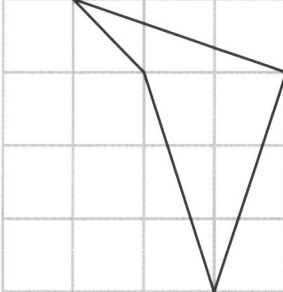

   Find the area of the shape shown by the black line.

   Answer _____ cm² [4]

3. Four congruent (identical) circles are placed inside a square of side 4 cm so that their circumferences touch the sides of the square and each other.

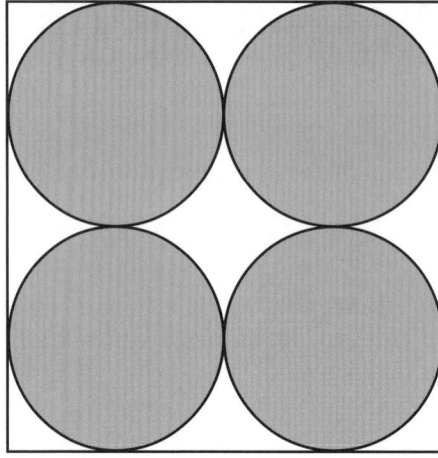

Find the area of the square which is outside the circles.

Answer _____ cm² [4]

4. Work out the value of:

$$\frac{1}{2} \times \frac{2}{3} \times \frac{3}{4} \times \frac{4}{5} \times \frac{5}{6} \times \frac{6}{7} \times \frac{7}{8} \times \frac{8}{9} \times \frac{9}{10} = \frac{}{}$$

Answer _____ [2]

5. Each of the boxes below has had a single digit removed from it.
   (a) Complete the multiplication problem by filling in the correct digits.

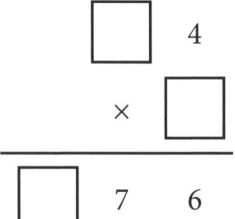

   (b) Complete a second solution to the problem above.

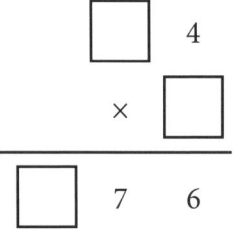

[6]

Answers

Revision Exercise 1A (non-calculator)

1. (a) 68 [A1] (b) Approximately 7 [A1]
2. (a) Evens [A1] (b) Certain [A1] (c) Unlikely [A1]
   (d) Likely [A1]
3. (a) £4 + £4 + £3 + £5 = £16 [M1A1] (b) Wrong because $9^2 = 81$ so the square root of 81 is 9 and 80 is very close to 81 [MA2]
4. 3954 (closer than 4359) [A2]
5. 1 five pence, 3 two pence, 4 one pence. [MA3]
6. (a) There may not be an equal number of boys and girls in the youth club. [MA1]
   (b) 30 ÷ (30 + 24) = ³⁰⁄₅₄ = ⁵⁄₉ [MA1A1]
7. (a) 25 × 2.2 = 55 (allow 1 mark for 50) [M1A1]
   (b) 140 × 1.6 = 160 [M1A1]
8. ⁵⁄₁₅ = ⅓ [M1A1]
9. (a) 7 × 2 = 14 [M1A1] (b) 30 ÷ 2 = 15 [M1A1]
   (c) Yes because 51 is an odd number and the patterns are all even numbers. [MA2]
10. Lines drawn 5 cm and 12 cm at right angles. [MA2]
    Hypotenuse measured as 13 cm. [A1]
11. (H, H) (H, T)
    (T, H) (T, T) [MA1]
    Probability of 2 heads = ¼ [MA1]
12. There are 3 triangles in the shape (see sketch below) which is 3 × 180° = 540° [MA1]
    115° + 96° + 143° + 81° = 435° [MA1]
    x = 540° − 435° = 105° [MA1]

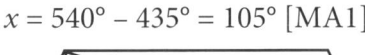

13. (a) Graph [MA2]

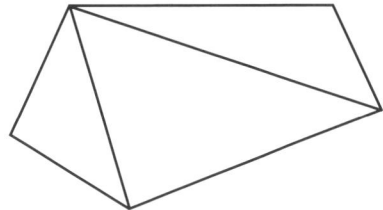

    (b) 0740 [A1] (c) 30 km [A1]
14. ½ + ⅓ = ⅚ [MA1]
    One sixth ate salmon = 16 people [MA1]
    Total attending = 16 × 6 = 96 people [M1A1]
15. (a) Coordinates of B are (−1, 1), (−2, 1) and (−1, 3) [MA1]. Coordinates of enlarged shape are (3, 2), (3, 6) and (5, 2) [A3]

Revision Exercise 1B (non-calculator)

1. (a) impossible [A1] (b) certain [A1] (c) evens [A1]
   (d) unlikely [A1]
2. (a) 5 [A1]
   (b) [A2]

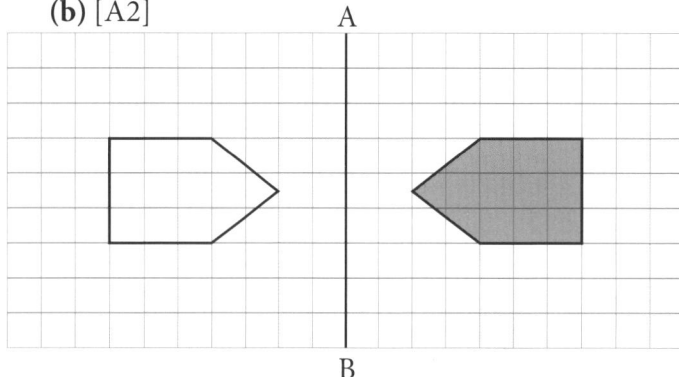

   (c) 2 [A1]
3. £5 − (4 × 49) = £5 − £1.96 = £3.04 [M1A1]
4. All marked on the scale: (a) X at ½ [A1]
   (b) Y at 0 [A1] (c) Z at ⅙ [A1]
5. (a) 8 × 100 = 800 [M1A1] (b) 27 ÷ 3 = 9 [M1A1]
6. (a) 1 − 0.8 = 0.2 [M1A1] (b) ²⁄₁₀ = ⅕ [MA1A1]
7. (a) 1 line of symmetry [A1] (b) M has coordinates (5, 0), (6, 0), (6, −2), (8, −2), (8, −3), (5, −3) [A1]
   (c) N has coordinates (1, −1), (1, −4), (2, −4), (2, −2), (4, −2), (4, −1) [A2]
8. (a) Impossible or zero [A1] (b) A [A1] (c) B [A1]
   (d) Probability of a 3 is ³⁄₆ = ½ so even chance [A1]
9. The first hour costs 2 × £5 = £10 [MA1]
   The second hour and each hour after that costs 2 × £2 = £4 [MA1]
   Two hours costs £10 + £4 = £14
   Three hours costs £14 + £4 = £18
   Four hours costs £18 + £4 = £22 [MA1]
   Since they have only £3 left over, they cannot hire the bicycles for another hour. Hence the answer is 4 hours. [MA1]
10. ⅓ + ¾ = ⁴⁄₁₂ + ⁹⁄₁₂ = ¹³⁄₁₂ or 1 ¹⁄₁₂ [MA2]

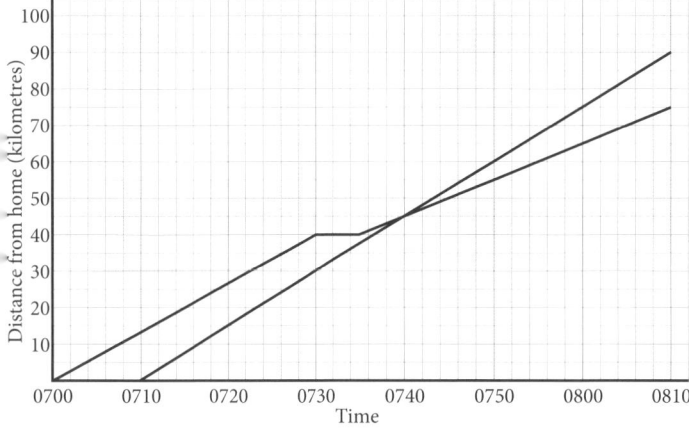

11. 6 ways (PQR), (PRQ), QPR, QRP, RPQ and RQP [A1A1 – 1 for each pair]
12. ¾ × 7 = ²¹⁄₄ = 5¼, so 6 tins [M1A1]
13. ¼ × 300 = 75 [M1A1]
14. (a) 26.5 (b) 21 (c) 44 pounds = 20 kilograms, 440 pounds = 200 kilograms [MA3]
15. (a) 40.232 [A1] (b) 5350 [A1]
16. Rotation of 90° clockwise about the origin (0, 0) [A1A1]

Revision Exercise 2A (with calculator)

1. 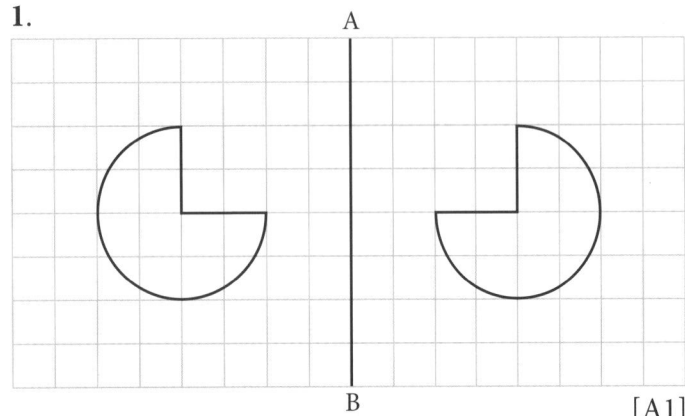 [A1]

2. (a) 1.5 × £4.20 = £6.30 [MA1].
   3 × £2.45 = £7.35 [MA1].
   Change = £20.00 – £6.30 – £7.35 = £6.35 [MA1]
   (b) 13 complete £s spent so 10 × 13 = 130 [MA1]
3. (a) unlikely [A1] (b) impossible [A1]
   (c) certain [A1] (d) likely [A1]
4. (a) [A1]
   (b) 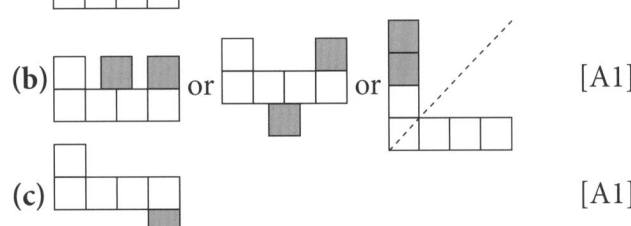 or or [A1]
   (c) [A1]
5. (a) Graph, points, line [A1A1]. (b) Approx 56 km [A1] (c) 60 miles = 96 km, so 120 miles = 2 × 96 = 192 km [MA2]. Or 100 miles = 160 km and 20 miles = 32 km, so 160 + 32 = 192 km
6. £2 each gives 6 × £2 = £12 [MA1]. 2 for price of 1 = £2 for 2 so £6 for 6 [MA1]. 1 free and pay for 2 = £4 so 6 costs 2 × £4 = £8 [MA1]. 40% discount means £1.20 each × 6 = £7.20 [MA1]. So "2 for the price of 1" is the best deal [A1].
7. (1, H), (2, H), (3, H), (4, H), (5, H) [A1]
   (1, T), (2, T), (3, T), (4, T), (5, T) [A1]

8. (a) Square [A1] (b)
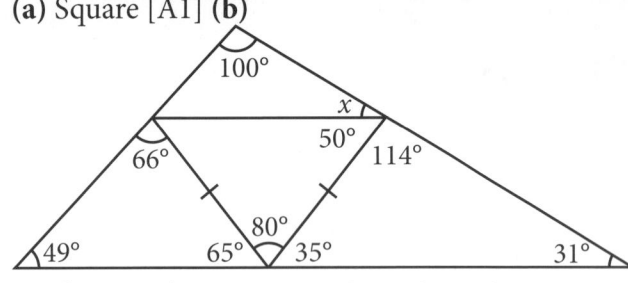
Find 65° angle as 180° – 66° – 49° = 65° (since the angles in a triangle add up to 180°)
Find 35° angle as 180° – 65° – 80° = 35° (since there are 180° in a straight line) [MA1]
Find the 114° angle as 180° – 35° – 31° = 114° (since there are 180° in a straight line)
Find the 50° angle as 180° – 80° = 100° (since the angles in a triangle add up to 180°) and the triangle is isosceles so both angles are equal = 50° [MA1]
So angle x = 16° as 180° – 114° – 50° = 16° (since there are 180° in a straight line) [MA1]
9. 6 inches = 15 cm; 1 yard = 36 inches so 1 yard = 15 cm × 6 (as 36 = 6 × 6) [M1] = 90 cm [A1]. 10 yards = 10 × 90 = 900 cm = 9 m [MA1].
10. (a) 4 × 3 × 2 × 1 = 24 [M1A1]
    (b) 4 is first digit so 3 choices for second, 2 for third, 1 for last = 3 × 2 × 1 = 6 [M1A1].
11. (a) $4x – 20 = 2x + 16$ [MA1]; $4x – 2x = 20 + 16$; $2x = 36$ [MA1]; $x = 18$ [MA1].
    $4x – 20 = 4(18) – 20 = 52$;
    or $2x + 16 = 2(16) + 16 = 52$ [MA1]
    (b) If ∠A = 52° then ∠C = 52° as opposite angles in a rhombus are equal. Sum of 2 other angles is 360 – 52 – 52 = 256 [MA1]
    Hence ∠B = 256 ÷ 2 = 128° [MA1].
12. (a) 1 – 0.35 – 0.25 – 0.23 = 0.17 [M1A1]
    (b) 0.17 × 600 = 102 [M1A1]
13. Mary : Elizabeth is 3 : 5 = 8 shares [M1]
    Elizabeth has 2 shares more than Mary = £24, so 1 share = £24 ÷ 2 = £12 [MA1] Mary gets 3 shares = £12 × 3 = £36, Elizabeth £12 × 5 = £60 and John gets £12 × 2 = £24 [M1A1]
    Total = £24 + £36 + £60 = £120 [A1].

Revision Exercise 2B (with calculator)

1. (a) Impossible or zero [A1] (b) ¾ [A1]
2. (a)

		Even numbers			
		2	4	6	8
	1	3	5	7	9
Factors of six	2	4	6	8	10
	3	5	7	9	11
	6	8	10	12	14

[MA4]